AF270234

# MY ARCTIC TERN
## MIGRATION JOURNEY

BY NANCY LOEWEN          ILLUSTRATED BY MARI LOBO

PICTURE WINDOW BOOKS
a capstone imprint

Published by Picture Window Books, an imprint of Capstone
1710 Roe Crest Drive, North Mankato, Minnesota 56003
capstonepub.com



Library of Congress Cataloging-in-Publication Data is available on the Library of Congress website.

ISBN: 9780756585372 (hardcover)
ISBN: 9780756585587 (paperback)
ISBN: 9780756585594 (ebook PDF)

Summary: Follow an Arctic tern on its magnificent migration journey.

Designer: Dina Her



Printed and bound in China. 6096

I'm a hungry arctic tern!

Hang on just a sec . . .

4

Excuse me while I finish my lunch. Mmmm, fish!
Did you see me **hover** and then dive in headfirst?
That's called plunge-diving. We arctic terns are great
at that.

It's late March, and I'm with my **colony** on pack ice around Antarctica. Here in the Southern **Hemisphere**, summer is nearly over. There is less daylight every day.

It's time for us to head north. We'll fly thousands of miles, all the way to the Arctic Circle.

We always follow the light. We need it
to see fish and other **prey** in the water.

Shhh. We're quiet now. You won't hear a chirp or a peep.
This moment of stillness before we leave is called a "dread."

And we're off!

We arctic terns are natural **gliders**. We're light and long. We weigh only about 4 ounces (113 grams). We're about 13 inches (33 centimeters) long. Our wingspan is a little over twice our length.

Hours pass. Every so often, we swoop down to the ocean to eat. But we mostly stay high in the sky. We can even sleep while gliding.

Did you know our colors change? When we are young, we have black legs and a black bill. We also have less black on our heads. When we are three or four years old, we become **breeding** adults. Then our legs and bills turn red during the breeding season.

Arctic terns usually live 20 to 30 years. I'm a seven-year-old male.

Weeks pass. Now we're flying over the **equator**. Halfway there!

We don't fly in a straight line. It's easier to fly with the wind. The wind patterns lead us across the ocean. We also go where the weather is good and the fish are plentiful.

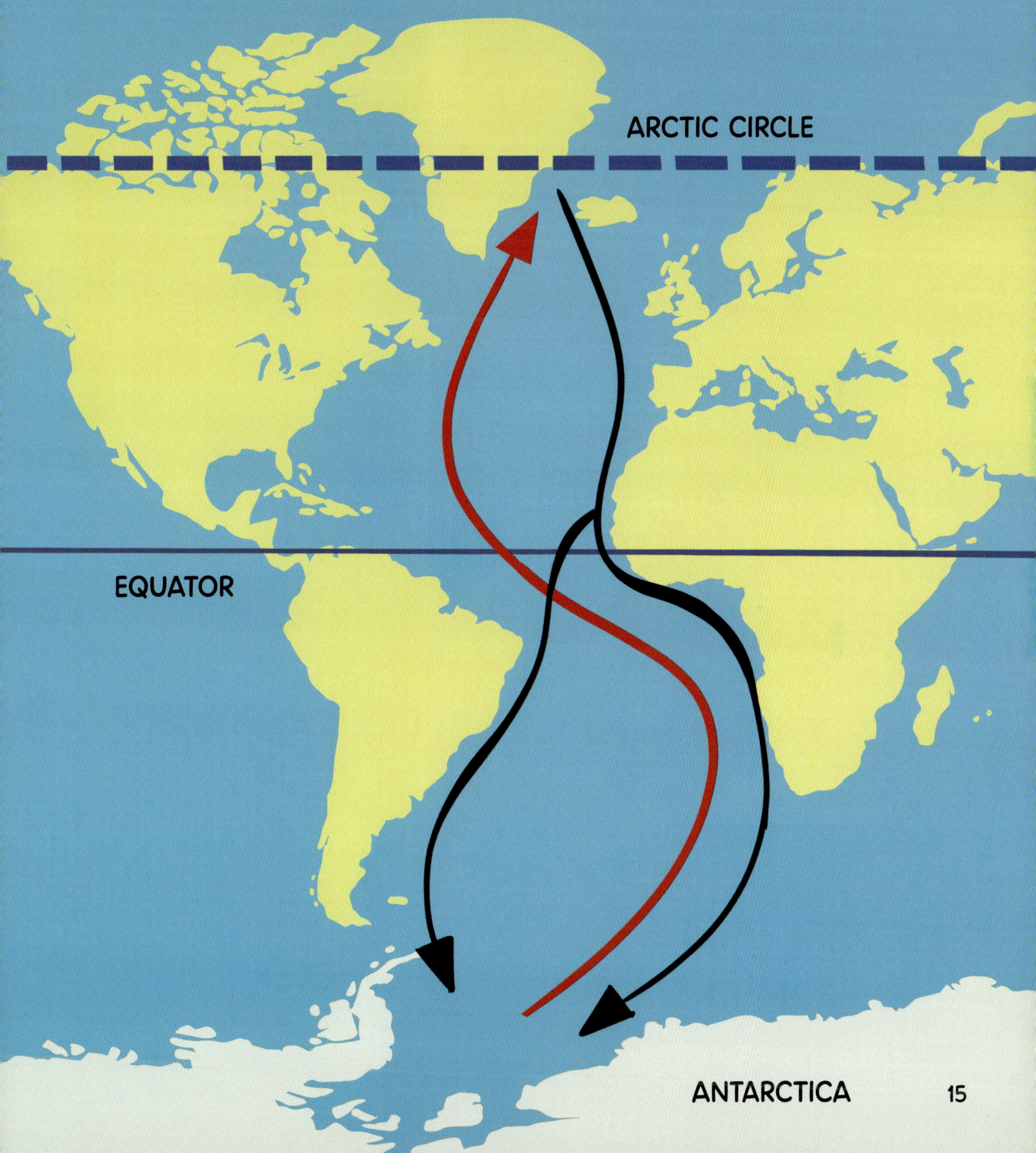

My colony has hundreds of birds, but we fly in smaller groups. I'm making the trip with four other terns.

The Atlantic Ocean is far below us. Its colors
are always changing. Weeks pass.

We're here! It's early June, so we're right on schedule. Our nesting grounds are in the **tundra** coastal area along the Arctic Ocean. Some colonies settle farther inland, near lakes and rivers.

For the next three months, this is home, sweet home! With long hours of daylight and plenty of fish, we're all set.

Meet my mate. Arctic terns usually mate for life. Sweet, right?

Our nests are simple. We find shallow dips in the ground.
Then we add pebbles or bits of moss around the edges.
My mate lays two eggs. We take turns keeping them warm.

Whoa! An arctic fox is trying to get our eggs. KLEEK! KLEEEK! I screech and dive at the fox until it slinks away.

21

Three weeks pass. The eggs move ever so slightly. Cracks appear.

Finally, here they are, our baby chicks! Their light orange beaks and legs will darken. Their feathers will turn white and black as they grow. My mate and I bring them fish to eat.

The chicks leave the nest in a couple of days. But they
don't go far. They hide in the rocks nearby.
In a few weeks, their flight feathers grow in. Then
they practice flying and catching fish. They're naturals!

It's early September. There are fewer hours of daylight. Time to head south! Even our little ones will make the trip. Here we are, back in the sky.

This time we're not in quite as much of a hurry. If
there's a good source of food, we might stop for a while.

Storm clouds are gathering in the distance. No worries!
We'll just fly around the storm.

Week after week, we follow the light.
In December, we reach Antarctica.

We eat and rest.
Some of our feathers fall off. This is called **molting**.
New feathers will grow in by the time we fly north again.

Every year, we fly up to 51,000 miles (82,000 kilometers)! That's the longest migration in the world. Over our lifetimes, many of us will fly more than 1.5 million miles (2,414,000 km). Along the way, we experience more daylight than any other living creature!

## ABOUT THE AUTHOR

Nancy Loewen grew up on a farm in southwestern Minnesota, surrounded by library books and cats. She now lives in Saint Paul and has published more than 140 children's books. Her Writer's Toolbox series received a Distinguished Achievement Award from the Association of Educational Publishers. Nancy has two adult children and a cat who sometimes nips at her knees under the table as she writes. Learn more at: nancyloewen.net.

## ABOUT THE ILLUSTRATOR

Mari Lobo is a children's book illustrator and toy designer, and her inner child couldn't be happier with her life choices! She was born in Sao Paulo, Brazil, and lives in California with her husband, daughter, and two dogs. She loves all animals and would pet a crocodile if it didn't bite her!

# GLOSSARY

**breed** (BREED)—to mate and produce young

**colony** (KAH-luh-nee)—a large group of the same kind of animal

**equator** (i-KWAY-tuhr)—an imaginary line around the middle of Earth

**glide** (GLIDE)—to move smoothly and easily with little effort

**hemisphere** (HEM-uhss-fihr)—one half of the Earth; the equator divides the Earth into northern and southern hemispheres

**hover** (HUHV-ur)—to remain in one place in the air

**molt** (MOHLT)—to shed fur, feathers, or an outer layer of skin; after molting, a new covering grows

**prey** (PRAY)—an animal hunted by another animal for food

**tundra** (TUHN-druh)—a cold area where trees do not grow; the soil under the ground in the tundra is permanently frozen

# INDEX